AF335330

Using Hard Problems to Create Pseudorandom Generators

ACM Distinguished Dissertations

1982
Abstraction Mechanisms and Language Design, Paul N. Hilfinger
Formal Specification of Interactive Graphics Programming Language, William R. Mallgren
Algorithmic Program Debugging, Ehud Y. Shapiro

1983
The Measurement of Visual Motion, Ellen Catherine Hildreth
Synthesis of Digital Designs from Recursion Equations, Steven D. Johnson

1984
Analytic Methods in the Analysis and Design of Number-Theoretic Algorithms, Eric Bach
Model-Based Image Matching Using Location, Henry S. Baird
A Geometric Investigation of Reach, James U. Korein

1985
Two Issues in Public-Key Cryptography, Ben-Zion Chor
The Connection Machine, W. Daniel Hillis

1986
All the Right Moves: A VLSI Architecture for Chess, Carl Ebeling
The Design and Evaluation of a High Performance Smalltalk System, David Michael Ungar

1987
Algorithm Animation, Marc H. Brown
The Rapid Evaluation of Potential Fields in Particle Systems, Leslie Greengard

1988
Computational Models of Games, Anne Condon
Trace Theory for Automatic Hierarchical Verification of Speed-Independent Circuits, David
L. Dill

1989
The Computational Complexity of Machine Learning, Michael J. Kearns
Uses of Randomness in Algorithms and Protocols, Joe Kilian

1990
Using Hard Problems to Create Pseudorandom Generators, Noam Nisan

Using Hard Problems to Create Pseudorandom Generators

Noam Nisan

The MIT Press
Cambridge, Massachusetts
London, England

This book was set in TeX by the author and was printed and bound in the United States of America.

Library of Congress Cataloging-in-Publication Data

Nisan, Noam
  Using hard problems to create pseudorandom generators / Noam Nisan.
    p.   cm. — (ACM distinguished dissertations)
  Revision of the author's thesis (Ph. D.)—University of California, Berkeley, 1988.
  Includes bibliographical references (p.   ).
  ISBN 0-262-14051-9
  1. Random number generators. 2. Computational complexity. I. Title. II. Series.
QA298.N57 1991
519.4—dc20   91-40779
                                                            CIP

Contents

Preface

This book is a slight revision of my doctoral dissertation, which was completed in December 1988 at the University of California at Berkeley under the supervision of Richard Karp. The subject of this thesis is pseudorandom generators, functions that stretch a short random string to a long string that "looks" random. This notion was introduced in the early 1980's by M. Blum, S. Micali, and A. Yao, who also observed that the construction of such pseudorandom generators is dependent upon the computational hardness of certain problems. Here we continue exploring the connection between computational hardness and pseudorandom generation and present two different constructions of pseudorandom generators, which are based upon the hardness of certain problems.

Our first construction is simple and very general. We construct pseudorandom generators that produce strings that look random to any algorithm from a complexity class C (e.g. P, NC, $PSPACE$, etc.) using an *arbitrary* function that is hard for C. This construction reveals an *equivalence* between the problems of proving certain lower bounds and of constructing pseudorandom generators.

This construction has many consequences. The most direct one is that efficient deterministic simulation of randomized algorithms is possible under much weaker assumptions than previously known. The efficiency of the simulations depends on the strength of the assumptions and may achieve $P = BPP$. We believe that our results are very strong evidence that the gap between randomized and deterministic complexity is not large.

Using the known lower bounds for constant depth circuits, our construction yields unconditionally proven pseudorandom generators for constant depth circuits. As an application we characterize the power of NP with a random oracle.

Our second pseudorandom generator produces strings that look random to all *Logspace* machines. This is proved without relying on any unproven assumptions. Instead, we use lower bounds on the complexity of the following multiparty communication game:

Let $f(x_1, ..., x_k)$ be a Boolean function that k parties wish to collaboratively evaluate. The i'th party knows each input argument except x_i, and each party has unlimited computational power. They share a blackboard, viewed by all parties, where they can exchange messages. The objective is to minimize the number of bits written on the board.

We prove lower bounds on the number of bits that need to be written on the board in order to compute a certain function. We then use these bounds to construct a pseudorandom generator for Logspace. As an application we present an explicit construction of universal traversal sequences for general graphs.

Acknowledgments

I would first like to thank my advisor, Richard Karp, who has helped me throughout my research, from his posing the first research problem I ever solved to his careful reading of my thesis. His clarity, impeccable teaching style, and scholarly attitude toward research will always be something to aspire to.

I am grateful to have had the privilege of working closely with Mike Sipser during the year he spent in Berkeley. It is hard to exaggerate the extent to which he has influenced my view of computer science.

While visiting the Hebrew University of Jerusalem, I had the pleasure of working with Avi Wigderson. The research described in chapter two was done with his help, guidance, and collaboration. I greatly appreciate his friendship and all that I have learned from him.

The results described in chapter three are part of my joint work with Laci Babai and Mario Szegedy at the University of Chicago. I thank them for their collaboration and hospitality.

I have learned much from the many discussions I have had with Manuel Blum and Umesh Vazirani, and from classes they have taught here. I was first introduced to pseudorandom generators in Manuel's Cryptography class, and I can trace the roots of the work described in chapter two back to a homework problem posed by Umesh in his class on Randomness.

One of the best things about studying theoretical computer science at Berkeley is the great group of students and visitors I have had the pleasure to meet and work with. They have been more than friends. Many, many research discussions with Steven Rudich and Russell Imagliazzo have strongly influenced my thinking. I have also enjoyed the opportunity of working with Amos Fiat, Rajeev Motwani, Moni Naor, and Danny Soroker. Ronitt Rubenfield deserves special thanks for her comments on this manuscript. I am grateful to all of them as well as to Lisa Hellerstein, Sampath Kannan, Valerie King, Assaf Shuster, David Zuckerman, and the rest of the theory crowd, for making the department what it is.

Finally, thanks, Cindy!

1 Introduction

1.1 Randomized Complexity

In the last decade randomization became an important tool in the design of algorithms. There are many problems for which efficient randomized algorithms have been given even though no efficient deterministic algorithm is known. This apparently enhanced power which randomization provides has become a major research topic in complexity theory.

For almost any natural complexity class, a corresponding randomized class may be defined (in fact several randomized variants can be defined). In most cases no nontrivial relationship is known between the corresponding randomized and deterministic classes, and in many cases there are some interesting problems known to lie in the randomized class but not known to lie in the corresponding deterministic one. Primality testing is known to lie in randomized polynomial time [SS77, Rab80, AH87], but not known to lie in deterministic polynomial time. Construction of a perfect matching in a graph is known to lie in random-NC [KUW86, MVV87], but not known to lie in NC. Undirected connectivity is known to lie in random-Logspace [AKL+79], but not known to lie in Logspace. Graph non-isomorphism is known to lie in AM, the randomized analogue of NP [GMW86], but not known to lie in NP.

This thesis continues the investigation into the power of randomization, and the relationships between randomized and deterministic complexity classes. The line of attack we pursue is the idea of *emulating* randomness, known as *pseudorandom generation*.

1.2 Pseudorandom Generators

The major conceptual idea behind pseudorandom generation is that sequences of events may *look* random even though they are not truly random in the information theoretic sense. Sequences may look random to any observer that does not have enough *computational* power to "understand" them.

This revolutionary idea was introduced and formalized in the early '80s by Blum and Micali [BM84], who were influenced by Shamir [Sha81], and by Yao [Yao82]. Blum and Micali and Yao proposed the idea of *pseudorandom generators*, functions which stretch a short string of truly random bits into a long string of bits which looks random to observers having limited computational power.

There are several motivations for research into pseudorandom generators. Perhaps the most broadly stated one is merely getting better insight into the nature of randomness from a *behavioral* point of view. More specifically, pseudorandom generation is perhaps the most general and natural method to reduce or eliminate the randomness required by algorithms. Pseudorandom sequences may replace the random sequences required by algorithms without changing the results in any way. This reduces the number of random bits required for solving the problem. Moreover, the trivial exponential deterministic simulation of randomized algorithms can now be used to obtain rather fast deterministic simulations.

Another motivation for research into pseudorandom generation is cryptography. A standard element implicit in many cryptographic protocols is to generate messages that do not give any information to your opponent, in other words, messages that appear random to him. Pseudorandom generators are a general framework to study these issues. Indeed many cryptographic protocols may be based on pseudorandom generators (see, e.g., [LR86, ILL89]).

Finally, perhaps the practical motivation should be mentioned: in reality many computer programs use random numbers. Hardly ever are truly random bits supplied by the computer (or do they even claim to be supplied by some physical means such as Zener diodes). Usually some pseudorandom generator supplies numbers which are hoped to be as good as truly random ones. It is important to know how much and in what cases this can be really justified.

1.3   Basic Definitions

Blum and Micali [BM84] and Yao [Yao82] were the first to define pseudorandom generators. They were concerned with pseudorandom generators that look random to polynomial time Turing machines, or to polynomial size circuits. We will consider the natural extensions and give general definitions of pseudorandom generators that look random to an arbitrary class of machines.

Another way in which we modify the definitions given by Blum-Micali and by Yao is the requirement regarding the running time of the pseudorandom generator. Blum-Micali and Yao defined pseudorandom generators to be computed in polynomial time. We will remove this requirement from the definition. Of course, for our pseudorandom generators to be interesting, we will need to show that they can indeed be computed somewhat efficiently.

The definitions that appear here are generic, and more precise specific definitions for particular complexity classes will appear where we use them.

Blum-Micali and Yao gave competing definitions of what a pseudorandom generator should be. These two definitions turned out to be equivalent. Yao's definition is perhaps the strongest one imaginable: that the pseudorandom bits will behave just like truly random ones in any way measurable by machines in the class.

Definition 1 A function $G = \{G_n : \{0,1\}^{m(n)} \to \{0,1\}^n\}$ is called a *pseudorandom* generator for class C if for every algorithm A in C, every polynomial $p(n)$, and for all sufficiently large n:

$$|Pr[A(y) = 1] - Pr[A(G_n(x)) = 1]| \le 1/p(n) \tag{1.3.1}$$

Where x is chosen uniformly at random in $\{0,1\}^{m(n)}$, and y in $\{0,1\}^n$.

The definition given by Blum-Micali seems to be a rather minimalist one: that given any prefix of the pseudorandom sequence, the next bit would look random.

Definition 2 A function $G = \{G_n : \{0,1\}^{m(n)} \to \{0,1\}^n\}$ passes all class C *prediction tests* if for every algorithm A in C, every $1 \le i \le n$, every polynomial $p(n)$, and all sufficiently large n:

$$|Pr[A(y_1, ... y_{i-1}) = y_i] - 1/2| \le 1/p(n) \tag{1.3.2}$$

Where y_j is the j'th bit output by G_n, and the probability is taken over a random input to G_n.

It was proven by Yao [Yao82] that these two definitions are equivalent.

THEOREM 1 (Yao) G is a pseudorandom generator for class C iff it passes all class C prediction tests.

This fact is extremely helpful, since the usual method of proving that a generator is pseudorandom is to show that it satisfies the weaker definition, and then to conclude that it has all the nice properties of the stronger one.

1.4 Previous Work

Most work regarding pseudorandom generators has been directed towards pseudorandom generators for P, polynomial time Turing machines. The first pseudorandom number generator was designed by Blum and Micali [BM84]. It is based on the unproven assumption that computation of the "discrete log" function cannot be done in polynomial time. They first showed that, under this assumption, the most significant bit of the discrete log cannot be approximated by any polynomial time computation, and they proceeded to give a general scheme to produce many pseudorandom bits using this fact.

Yao [Yao82] generalized this construction. He showed how any *one-way permutation* can be used in place of the discrete log function.

Definition 3 A function $f = \{f_n : \{0,1\}^n \to \{0,1\}^n\}$ is called a one-way permutation if (1) For all n, f_n is one-one and onto (2) f can be computed in polynomial time (3) Any polynomial size circuit that attempts to invert f errs on at least a polynomially large fraction of the inputs.

Yao first showed how to "amplify" the "unpredictability" of condition 3 and obtain a "hard bit", a bit that cannot be predicted at all. This amplification is achieved by taking multiple copies of the hard function on disjoint sets of input bits and xor-ing them; Yao shows that this operation indeed "amplifies" the unpredictability of the bits. Once a "hard bit" is obtained, a pseudorandom generator can be designed using the Blum-Micali scheme:

A seed x_0 of size n^ϵ bits is given as the random input. The one-way permutation is applied to the seed repeatedly, generating a sequence $x_0, f(x_0), f(f(x_0)), \ldots, f^{(n)}(x_0)$. The pseudorandom output of the generator is obtained by extracting the "hard bit" from each string in this sequence. The proof that this is indeed a pseudorandom generator proceeds by showing how a test that this sequence fails can be used to invert the one-way permutation f.

THEOREM 2 (Yao) If a one-way permutation exists then for every $\epsilon > 0$ there exists a polynomial time computable pseudorandom generator $G : \{0,1\}^{n^\epsilon} \to \{0,1\}^n$.

Recently, Impagliazzo, Levin and Luby [ILL89] and Hastad [Has90] proved that the existence of any one-way function (not necessarily a permutation) suffices for the construction of pseudorandom generators. This condition is also necessary for the existence of pseudorandom generators that can be computed in polynomial time. The pseudorandom generators described so far can indeed be computed in polynomial time.

All the work described so far concerns generators that pass all *polynomial time* tests; pseudorandom generators for P. Some work has also been done regarding the construction of pseudorandom generators for other complexity classes.

Reif and Tygar [RT84] describe a generator that passes all NC tests and, moreover, can itself be computed in NC. This generator is based on the assumption that "inverse mod p" cannot be computed in NC. The main innovation here is showing that for this particular function, the original Blum-Micali-Yao generator can be parallelized.

Ajtai and Wigderson [AW85] considered pseudorandom generators for AC^0. They use an ad-hoc construction based on the lower bound methods for constant depth circuits to construct a pseudorandom generator that passes all AC^0 tests. The significance of this result is that it does not require any unproven assumptions. Instead, it builds upon the known, proven lower bounds for constant depth circuits.

1.5 Hardness vs. Randomness

The second chapter of this thesis is devoted to a new general construction of pseudorandom generators for arbitrary complexity classes. This construction overcomes two basic limitations of the known pseudorandom generators:

- They require a strong unproven assumption. (The existence of a one-way function, an assumption which is even stronger than $P \neq NP$.)
- They are sequential, and cannot be applied to an arbitrary complexity class. E.g. there is no known construction of pseudorandom generators for NC that is based upon a general complexity assumption about NC.

The construction which we propose avoids both problems: It can be applied to any complexity class C, it is based on an arbitrary function which is hard for C, and gives a pseudorandom generator that looks random to the class C. Although our generator cannot necessarily be computed in the class C, we will show that it can be computed efficiently enough for our *simulation* purposes.

Perhaps the most important conceptual implication of this construction is that it proves the *equivalence* between the problem of proving lower bounds for the size of circuits approximating functions in EXPTIME and the problem of constructing pseudorandom generators which run "sufficiently fast". This should be contrasted with the results of Impagliazzo, Levin, Luby and Hastad [ILL89, Has90] showing the equivalence of proving the existence of *one-way* functions and constructing pseudorandom generators that run in polynomial time. Our construction requires much weaker assumptions but yields less efficient pseudorandom generators. This loss does not have any effect when using pseudorandom generators for the deterministic simulation of randomized algorithms.

This construction has many implications, and a large part of the second chapter describes them. We first show that efficient deterministic simulation of randomized algorithms is possible under much weaker assumptions than previously known. The efficiency of the simulation depends on the strength of the assumption and can be good enough to show $P = BPP$. Since the assumptions required for our generator are so weak and natural, we believe that this work provides overwhelming evidence that the gap between deterministic and randomized complexity is not large.

We then turn to pseudorandom generators for constant depth circuits. Since lower bounds for constant depth circuits are known (e.g., [Has86]), our construction yields an unconditionally proven pseudorandom generator for constant depth circuits. This generator improves upon the known generator, due to Ajtai and Wigderson [AW85], and implies much better deterministic simulation of randomized constant depth circuits.

Our generator for constant depth circuits turns out to have some interesting consequences regarding the power of random oracles for complexity classes in the polynomial time hierarchy. We show that NP with a random oracle is exactly the class AM, solving an open problem of Babai [BM88]. We also show that a random oracle does not add power to the polynomial time hierarchy.

A new proof is given of the fact that BPP is in the polynomial time hierarchy. Our final application is a surprising connection between simulation of "time by space" and simulation of "randomness by determinism". We show that one of these simulations can be substantially improved over known simulations.

The results in this chapter have appeared in [Nis91a, NW88] and are joint work with Avi Wigderson.

1.6   Pseudorandom Generators for Logspace

The third chapter in this thesis describes the construction of a pseudorandom generator for Logspace. This result is unique in that it is not based on any unproven assumptions. The only other class for which pseudorandom generators were unconditionally proven to exist is AC^0.

In order to prove the correctness of our pseudorandom generator we use the following multiparty communication game, first introduced by Chandra, Furst and Lipton [CFL83]: Let $f(x_1 \ldots x_k)$ be a Boolean function that accepts k arguments each n bits long. There are k parties who wish to collaboratively evaluate f; the i'th party knows each input argument except x_i; and each party has unlimited computational power. They share a blackboard, viewed by all parties, where they can exchange messages. The objective is to minimize the number of bits written on the board.

We first prove lower bounds for the number of bits that must be written on the board in order to get even a small advantage on computing certain functions. We then show how to use these lower bounds in order to construct a pseudorandom generator for Logspace.

We conclude by giving some applications of our pseudorandom generator. We describe a construction of universal sequences for arbitrary regular graphs; no nontrivial such construction was previously known. We also show that random Logspace machines with two-way access to the random bits are better, in some specific sense, than random Logspace machines with the usual one-way access to the random bits.

The results in this chapter have appeared in [BNS89] and are joint work with Laszlo Babai and Mario Szegedy.

1.7 Recent Results

Since this thesis was originally written (in December 1989), several new results related to problems considered here were obtained. The most relevant ones are a different, improved construction for pseudorandom generators for Logspace [Nis90] and a further sharpening of the conditions under which the pseudorandom generators presented in chapter two can be constructed [BFNW91].

The fourth chapter briefly describes these as well as other recent results related to the subject of this thesis.

2 Hardness vs. Randomness

In this chapter we describe a general construction of pseudorandom generators which can be based on the hardness of an arbitrary function, and can be applied to any complexity class.

In section one we define, describe in detail, and prove the correctness of our pseudorandom generator. In section two we then prove a host of applications and corollaries of our construction.

2.1 The Generator

In this section we state and prove our results for pseudorandom generators that look random to small circuits, and thus also to time-bounded Turing machines. All the definitions and theorems we give have natural analogues regarding pseudorandom generators for other complexity classes such as depth-bounded circuits, etc. It is rather straightforward to make the required changes, and we leave it to the interested reader.

2.1.1 Definitions

Informally speaking, a pseudorandom generator is an "easy to compute" function which converts a "few" random bits to "many" pseudorandom bits that "look random" to any "small" circuit. Each one of the quoted words is really a parameter, and we may get pseudorandom generators of different qualities according to the choice of parameter. For example, the standard definitions are: "easy to compute" = polynomial time; "few" = n^ϵ; "many" = n; "look random" = subpolynomial difference in acceptance probability; and "small" = any polynomial. We wish to present a more general tradeoff and obtain slightly sharper results than these particular choices of parameters allow. Although all these parameters can be freely traded-off by our results, it will be extremely messy to state everything in its full generality. We will thus restrict ourselves to two parameters that will have multiple purposes. The choice was made to be most natural from the "simulation of randomized algorithms" point of view.

The first parameter we have is "the quality of the output". This will refer to 3 things: the number of bits produced by the generator, the maximum size of the circuit the generator "fools", and the reciprocal of difference in accepting probability allowed. In general, in order to simulate a certain randomized algorithm, we will require a generator with quality of output being approximately the running time of the algorithm.

The second parameter is "the price" of the generator. This will refer to both the number of input bits needed and to the logarithm of the running time of the generator. In general, the deterministic time required for simulation will be exponential in the "price" of the generator.

Definition 4 A function $G = \{G_n : \{0,1\}^{l(n)} \to \{0,1\}^n\}$, denoted by $G : l \to n$, is called a *pseudorandom generator* if for all n and any circuit C of size at most n:

$$|Pr[C(y) = 1] - Pr[C(G(x)) = 1]| \leq 1/n \qquad (2.1.1)$$

where y is chosen uniformly in $\{0,1\}^n$, and x in $\{0,1\}^{l(n)}$.

We say G is a *quick* pseudorandom generator if it runs in deterministic time exponential in its *input* size, $G \in DTIME(2^{O(l(n))})$.

The parameter n is the "quality" of the generator, and for quick generators, $l(n)$ is the "price".

We will also define an *extender*, a pseudorandom generator that only generates one extra bit:

Definition 5 : A function $G = \{G_l : \{0,1\}^l \to \{0,1\}^{l+1}\}$ is called an $n(l)$-*extender* if for all l and any circuit C, of size at most $n = n(l)$:

$$|Pr[C(y) = 1] - Pr[C(G(x)) = 1]| \leq 1/n \qquad (2.1.2)$$

where y is chosen uniformly in $\{0,1\}^{l+1}$, and x in $\{0,1\}^l$.

We say G is a *quick* extender if it runs in deterministic time exponential in its *input* size, $G \in DTIME(2^{O(l)})$.

The major difference between our definition and the "normal" definition is the requirement regarding the running time of the algorithm: normally the pseudorandom generator is required to run in polynomial time; we allow it to run in time exponential in its *input* size. This relaxation allows us to construct pseudorandom generators under much weaker conditions than the ones required for polynomial time pseudorandom generators, but our pseudorandom generators are as good for the purpose of simulating randomized algorithms as polynomial time ones. The following lemma is the natural generalization of Yao's [Y2] lemma showing how to use pseudorandom generators to simulate randomized algorithms:

LEMMA 1 If there exists a quick pseudorandom generator $G : l(n) \to n$ then for any time bound $t = t(n)$: $BPTIME(t) \subseteq DTIME(2^{O(l(t^2))})$. (Here $BPTIME(t)$ is the class of languages accepted with bounded two-sided error by randomized Turing machines running in time $O(t)$).

Proof: The simulation can be partitioned into two stages. First, the original randomized algorithm which uses $O(t)$ random bits is simulated by a randomized algorithm which uses $l(t^2)$ random bits but runs in time $2^{O(l(t^2))}$. This is done simply by feeding the original algorithm pseudorandom sequences obtained by the generator instead of truly random bits. Since the output of the pseudorandom generator looks random to any circuit of size t^2, and since any algorithm running in time t can be simulated by a circuit of size t^2, the output of the generator will look random to the original algorithm. Thus the probability of acceptance of this randomized algorithm will be almost (i.e. to within $1/n$) the same as that of the original one.

In the second stage we simulate this randomized algorithm deterministically, by trying all the possible random seeds and taking a majority vote. The number of different seeds is $2^{l(t^2)}$, and for each one a computation of complexity $2^{O(l(t^2))}$ is done. $\square$

2.1.2   Hardness

The assumption under which we construct a generator is the existence of a "hard" function. By "hard" we need not only that the function can not be computed by small circuits but also that it can not be *approximated* by small circuits. There are basically two parameters to consider: the size of the circuit and the closeness of approximation.

Definition 6 : Let $f : \{0,1\}^n \to \{0,1\}$ be a boolean function. We say that f is (ϵ, S)-*hard* if for any circuit C of size at most S,

$$|Pr[C(x) = f(x)] - 1/2| \leq \epsilon/2 \tag{2.1.3}$$

Where x is chosen uniformly at random in $\{0,1\}^n$.

Yao [Y2] shows how the closeness of approximation can be "amplified" by xor-ing multiple copies of f.

LEMMA 2 (Yao) Let $f_1, ..., f_k$ all be (ϵ, S)-hard, and $x_1, ..., x_k$ be disjoints sets of input bits. Then for any $\delta > 0$, the function $f_1(x_1) \oplus ... \oplus f_k(x_k)$ is $(\epsilon^k + \delta, \delta^2(1-\epsilon)^2 S)$-hard.

This lemma allows us to base our results on a function that has the following weak hardness properties.

Definition 7 Let $f = \{f_m : \{0,1\}^m \to \{0,1\}\}$ be a boolean function. We say that f *cannot be approximated* by circuits of size $s(m)$ if for some constant k, all large enough m, and all circuits C_m of size at most $s(m)$:

$$Pr[C_m(x) \neq f(x)] > m^{-k} \tag{2.1.4}$$

where x is chosen uniformly in $\{0,1\}^m$.

This is a rather weak requirement, as it only requires that small circuits attempting to compute f have a non-negligible fraction of error. Yao's xor-lemma allows amplification of such hardness to the sort of hardness which we will actually use in our construction. We will want that no small circuit can get any non-negligible advantage in computing f.

Definition 8 : Let $f = \{f_m : \{0,1\}^m \to \{0,1\}\}$ be a boolean function. The *Hardness* of f at m, $H_f(m)$ is defined to be the maximum integer h_m such that f_m is $(1/h_m, h_m)$-hard.

The following corollary is an immediate application of Yao's lemma.

COROLLARY 1 : Let $s(m)$ be any function (size-bound) such that $m \leq s(m) \leq 2^m$. If there exists a function f in EXPTIME that cannot be approximated by circuits of size $s(m)$, then for some $c > 0$ there exists a function f' in EXPTIME that has hardness $H_{f'}(m) \geq s(m^c)$.

2.1.3 The Main Lemma

Given a "hard" function, it is intuitively easy to generate one pseudorandom bit from it since the value of the function must look random to any small circuit. The problem is to generate more than one pseudorandom bit. In order to do this we will compute the function on many different, nearly disjoint subsets of bits.

Definition 9 A collection of sets $\{S_1, ..., S_n\}$, where $S_i \subset \{1, ..., l\}$, is called a (k, m)-*design* if:

- For all i: $|S_i| = m$
- For all $i \neq j$: $|S_i \cap S_j| \leq k$

A $n \times l$ 0-1 matrix is called a (k, m)-design if its n rows, interpreted as subsets of $\{1, ..., l\}$ are a (k, m)-design.

Definition 10 Let A be a $n \times l$ 0-1 matrix, let f be a boolean function, and let $x = (x_1, ..., x_l)$ be a boolean string. Denote by $f_A(x)$ the n bit vector of bits computed by applying the function f to the subsets of the x's denoted by the n different rows of A.

Our generator expands the seed x to the pseudorandom string $f_A(x)$. The quality of the bits is assured by the following lemma.

LEMMA 3 Let m, n, l be integers; let f be a boolean function, $f : \{0,1\}^m \to \{0,1\}$, such that $H_f(m) \geq n^2$; and let A be a boolean $n \times l$ matrix which is a $(\log n, m)$ design. Then $G : l \to n$ given by $G(x) = f_A(x)$ is a pseudorandom generator.

Proof: We will assume that G is not a pseudorandom generator and derive a contradiction to the hardness assumption. If G is not a pseudorandom generator then for some circuit C, of size n,

$$Pr[C(y) = 1] - Pr[C(G(x)) = 1] > 1/n \tag{2.1.5}$$

where x is chosen uniformly in $\{0,1\}^l$ and y in $\{0,1\}^n$. We first show, as in [GM84, Yao82], that this implies that one of the bits of $f_A(x)$ can be predicted from the previous ones.

For any i, $0 \leq i \leq n$, we define a distribution E_i on $\{0,1\}^n$ as follows: the first i bits are chosen to be the first i bits of $f_A(x)$, where x is chosen uniformly in $\{0,1\}^l$, and the other $n - i$ bits are chosen uniformly at random. Define

$$p_i = Pr[C(z) = 1] \tag{2.1.6}$$

where z is chosen according to the distribution E_i. Since $p_0 - p_n > 1/n$, it is clear that for some i, $p_{i-1} - p_i > 1/n^2$. Using this fact we will build a circuit that predicts the i'th bit.

Define a circuit D, which takes as input the first $i - 1$ bits of $f_A(x)$, $y_1, ..., y_{i-1}$, and predicts the i'th bit, y_i. D is a probabilistic circuit. It first flips $n - i + 1$ random bits, $r_i, ..., r_n$. On input $y =< y_1, ..., y_{i-1} >$, it computes $C(y_1, ..., y_{i-1}, r_i, ..., r_n)$. If this evaluates to 1 then D will return r_i as the answer, otherwise it will return the complement of r_i. As in [Yao82] it can be shown that D is correct with bias of at least $p_{i-1} - p_i$, i.e.

$$Pr[D_n(y_1, ..., y_{i-1}) = y_i] - 1/2 > 1/n^2 \tag{2.1.7}$$

where the probability is taken over all choices of x and of the random bits that D uses. At this point an averaging argument shows that it is possible to set the private random bits that D uses to constants and achieve a deterministic circuit D' while preserving the bias.

By now we have constructed a circuit that predicts y_i from the bits $y_1, ..., y_{i-1}$. To achieve a contradiction to the hardness assumption we will now transform this circuit to a circuit that predicts y_i from the bits $x_1, ..., x_l$. W.l.o.g. we can assume that y_i depends on $x_1, ..., x_m$, i.e.

$$y_i = f(x_1, ..., x_m) \tag{2.1.8}$$

Since y_i does not depend on the other bits of x, it is possible to set the other bits to constants, while leaving the prediction of y_i valid. By an averaging argument there exist constants $c_{m+1}, ..., c_l$ such that setting $x_j = c_j$ for all $m < j \leq l$ preserves the prediction probability. At this point, however, each one of the bits $y_1, ..., y_{i-1}$ depends only on at most $\log n$ of the bits $x_1, ..., x_m$. This is so since the intersection of the set of $x'_k s$ defined by y_i and by y_j is bounded from above by $\log n$ for each $i \neq j$. Now we can compute each y_i as a CNF (or DNF) formula of a linear (in n) size over the bits it uses. This gives us a circuit $D''(x_1, ...x_m)$ that predicts y_i which is $f(x_1, ..., x_m)$. It is easy to check that the size of D'' is at most n^2, and the bias achieved is more than $1/n^2$, which contradicts the assumption that $H_f(m) > n^2$. $\square$

2.1.4 Construction of Nearly Disjoint Sets

This section describes the actual construction of the designs that are used by the pseudo-random generator. In the construction of the generator, we are given a "hard" function f with a certain "hardness", H_f, and we wish to use it to construct a pseudorandom generator $G : l \rightarrow n$. Our aim is to minimize l, that is, to get a pseudorandom generator that uses the smallest number of random bits. If we look at the requirements of lemma 3, we see that we require a $(\log n, m)$-design, where m must satisfy $H_f(m) \geq n^2$. This basically determines a minimum possible value for m. The following lemma shows that l need not be much larger than m.

LEMMA 4 For every integers n and m, such that $\log n \leq m \leq n$, there exists an $n \times l$ matrix which is a $(\log n, m)$-design, where $l = O(m^2)$. Moreover, such a matrix can be computed by a Turing machine running in space $O(\log n)$.

Proof: We need to construct n different subsets of $\{1, ..., l\}$ of size m with small intersections. Assume, wlog, that m is a prime power, and let $l = m^2$. (If m is not a prime power, pick, e.g., the smallest power of 2 which is greater than m; this can at most double the value of m.) Consider the numbers in the range $\{1, ..., l\}$ as pairs of elements in $GF(m)$, i.e., we construct subsets of $\{< a, b > |a, b \in GF(m)\}$. Given any polynomial q on $GF(m)$, we define a set $S_q = \{< a, q(a) > |a \in GF(m)\}$. The sets we take are all of this form, where q ranges over polynomials of degree at most $\log n$. The following facts can now be easily verified:

- The size of each set is exactly m.
- Any two sets intersect in at most $\log n$ points. (Since two polynomials of degree $\log n$ can intersect on at most $\log n$ points.)
- There are at least n different sets. (Since the number of polynomials over $GF(m)$ of degree at most $logn$ is $m^{logn+1} \geq n$).

It should be noted that all that is needed to construct these sets effectively is simple arithmetic in $GF(m)$, and since m has a length of $O(\log n)$ bits, everything can be easily computed by a log-space bounded Turing machine. $\square$

It can be shown that the previous design is optimal up to a factor of $\log n$, i.e, for given m and n, the parameter l is within a log factor of the design with the smallest value of l. For most values of m this small added factor is not so important; however, for small values of m we may wish to do better. One way to achieve a better design for small values of m is to consider multivariate polynomials over finite fields. These multinomials may define sets in a similar manner as in the previous design, and for small values of m, l can be reduced up to about $m \log m$. We leave the details to the interested reader.

A case of special interest is $m = O(\log n)$. In this case it is possible to reduce l also to $O(\log n)$. We do not have an explicit construction for this; however, we note that such a design can be computed in polynomial time.

LEMMA 5 For every integers n and m, where $m = C \log n$ for some constant C , there exists a $n \times l$ matrix which is a $(\log n, m)$-design where $l = 2C^2 \log n$. Moreover, the matrix can be computed by a Turing machine running in time polynomial in n.

Proof: The Turing machine will *greedily* choose subsets of $\{1, ..., l\}$ of cardinality m, which intersect each of the previously chosen sets at less than $\log n$ points. A simple counting argument shows that it is always possible to choose such a set, whatever the previous sets that were chosen are, as long as there are at most n such sets. The running time is polynomial since we are looking at subsets of $O(\log n)$ elements. $\square$

2.1.5   Main Theorem

The main theorem we get is a necessary and sufficient condition for the existence of quick pseudorandom generators.

THEOREM 3 For every function (size bound) $l \leq s(l) \leq 2^l$ the following are equivalent:

1. For some $c > 0$ some function in EXPTIME cannot be approximated by circuits of size $s(l^c)$.
2. For some $c > 0$ there exists a function in EXPTIME with hardness $s(l^c)$.
3. For some $c > 0$ there exists a quick $s(l^c)$-*extender* $G : l \to l + 1$.
4. For some $c > 0$ there exists a quick pseudorandom generator $G : l \to s(l^c)$.

Proof: We will show $(1) \to (2) \to (4) \to (3) \to (1)$.
$(1) \to (2)$ is corollary 1.
$(4) \to (3)$ is trivial.
$(3) \to (1)$ is proven by the following observation: Let $G = \{G_l\}$ be an extender as in (3). Consider the problem of "Is y in the range of G?". It can be easily seen that this can be computed in exponential time; however, no circuit of size $s(l^c)$ can approximate it

since that circuit would distinguish between the output of G and between truly random strings.

$(2) \rightarrow (4)$. This is, of course, the main part. Let f be a function in EXPTIME with hardness $s(l^c)$. We build a quick pseudorandom generator $G : l \rightarrow n$, for $n = s(m^{c/4})$: For every n let A_n be the matrix guaranteed by lemma 4 for $m = \sqrt{l}$. Notice that this is an $n \times l$ matrix which is a $(logn, m)$-design. Notice also that, by our choice of parameters, $H_f(m) > n^2$. Thus, by lemma 3, the function $G_n(x) = f_{A_n}(x)$ is a pseudorandom generator. $G = \{G_n\}$ is a quick pseudorandom generator simply since f is in EXPTIME. $\square$

This theorem should be contrasted with the known results regarding the conditions under which *polynomial time* computable pseudorandom generators exist. Impagliazzo, Levin, Luby and Hastad [ILL89, Has90] prove the following theorem:

THEOREM 4 The following are equivalent (for any $1 > \epsilon > 0$):

- There exists a 1-way function.
- There exists a polynomial time computable pseudorandom generator $G : n^\epsilon \rightarrow n$.

The existence of polynomial time computable pseudorandom generators seems to be a stronger statement, and requires apparently stronger assumptions than the existence of "quick" pseudorandom generators.

2.2 Main Corollaries

2.2.1 Sequential Computation

The major application of the generator is to allow better deterministic simulation of randomized algorithms. We now state the results we get regarding the deterministic simulation of algorithms.

THEOREM 5 If there exists a function computable in $DTIME(2^{O(n)})$,

1. that cannot be approximated by polynomial size circuits. Or,
2. that cannot be approximated by circuits of size 2^{n^ϵ} for some $\epsilon > 0$. Or,
3. with hardness $2^{\epsilon n}$ for some $\epsilon > 0$.

Then

1. $BPP \subseteq \bigcap_{\epsilon > 0} DTIME(2^{n^\epsilon})$.
2. $BPP \subseteq DTIME(2^{(\log n)^c})$ for some constant c.

3. $BPP = P$.

respectively.

Proof: Using theorem 3, (1) implies the existence of a quick pseudorandom generator G : $n^\epsilon \to n$ for every $\epsilon > 0$, and (2) implies the existence of a quick pseudorandom generator $G : (\log n)^c \to n$ for some $c > 0$. (3) implies the existence of a quick pseudorandom generator $G : C \log n \to n$ for some $C > 0$. This can be seen by modifying the proof of theorem 3 as to use the design specified in lemma 5 instead of the "generic" design (lemma 4). The simulation results follow by lemma 1. $\square$

2.2.2 Parallel Computation

The construction of the generator was very general; it only depended on the existence of a function that was hard for the class the generator is intended for. Thus we can get similar simulation results for other complexity classes under the analogous assumptions. We will now state the major simulation results we get for parallel computation.

THEOREM 6 If there exists a function in PSPACE that

1. cannot be approximated by NC circuits. Or
2. cannot be approximated by circuits of *depth* n^ϵ (for some constant $\epsilon > 0$).

Then

1. $RNC \subset \bigcap_{\epsilon > 0} DSPACE(n^\epsilon)$.
2. $RNC \subset DSPACE(polylog)$.

Respectively.

Proof: The proof is the straightforward adaptation of our pseudorandom generator to the parallel case. The important point is that the generator itself is parallel, and indeed in the proof of the main lemma, the depth of the circuit C increases only slightly. $\square$

2.2.3 Constant Depth Circuits

A special case of interest is the class of constant depth circuits. Since for this class lower bounds are known, we can use our construction to obtain pseudorandom generators for constant depth circuits that do not require any unproven assumption.

Our generator is based upon the known lower bounds for constant depth circuits computing the parity function. We will use directly the strongest bounds due to Hastad [Has86].

THEOREM 7 (Hastad) For any family $\{C_n\}$ of circuits of depth d and size at most $2^{n^{1/(d+1)}}$, and for all large enough n:

$$|Pr[C_n(x) = parity(x)] - 1/2| \leq 2^{n^{1/(d+1)}} \qquad (2.2.9)$$

where x is chosen uniformly over all n-bit strings.

Applying to this our construction we get:

THEOREM 8 For any integer d, there exists a family of functions: $\{G_n : \{0,1\}^l \to \{0,1\}^n\}$, where $l = O((\log n)^{2d+6})$ such that:

- The family $\{G_n\}$ can be computed by a log-space uniform family of circuits of polynomial size and $d+4$ depth.
- For any family $\{C_n\}$ of circuits of polynomial size and depth d, for any polynomial $p(n)$, and for all large enough n: $|Pr[C_n(y) = 1] - Pr[C_n(G_n(x)) = 1]| \leq 1/p(n)$, where y is chosen uniformly in $\{0,1\}^n$, and x is chosen uniformly in $\{0,1\}^l$.

Proof: Again $G_n = f_{A_n}$ where f is the parity function, and A_n is the design described in lemma 4 for $m = (\log n)^{d+3}$. Notice that: (1) the generator can be computed by polynomial size circuits of depth $d+4$ since it is just the parity of sets of bits of cardinality $(\log n)^{d+3}$. (2) All the considerations in the proof of correctness of the generator apply also to constant depth circuits. In particular the depth of the circuit C in the proof of lemma 3 increases only by one. $\square$

We can now state the simulation results we get for randomized constant depth circuits. Denote by RAC^0 ($BPAC^0$) the set of languages that can be recognized by a uniform family of *Probabilistic* constant depth, polynomial size circuits, with 1-sided error (2-sided error bounded away from 1/2 by some polynomially small fraction).

THEOREM 9 $RAC^0 \subseteq BPAC^0 \subseteq \bigcup_c DSPACE((\log n)^c) \subseteq \bigcup_c DTIME(2^{(\log n)^c})$

Denote by $\#DNF$ the problem of counting the number of satisfying assignments to a DNF formula, and by $Approx - \#DNF$ the problem of computing a number which is within a factor of 2 from the correct value. Clearly $\#DNF$ is $\#P$ complete. However, our results imply that:

COROLLARY 2 $Approx - \#DNF \in DTIME(2^{(\log n)^{14}})$

Proof: Karp and Luby [KL83] give a probabilistic algorithm for $Approx - \#DNF$. It is not difficult to see that this algorithm can be implemented by RAC^0 circuits of depth 4. $\square$

2.2.4 Random Oracles

The existence of our pseudorandom generator for constant depth circuits has implications concerning the power of random oracles for classes in the polynomial time hierarchy.

Let C be any complexity class (e.g. P, NP, ...). As in [BM88] we define the class $almost - C$ to be the set of languages L such that:

$$Pr[L \in C^A] = 1 \qquad\qquad (2.2.10)$$

where A is an oracle chosen at random. The class $almost - C$ can be thought of as a natural probabilistic analogue of the class C.

The following theorem is well known [Kur87, BG81], and underscores the importance of BPP as the random analogue of P:

THEOREM 10 $BPP = almost - P$

Babai [Bab85] introduced the class AM. An AM Turing machine is a machine that may use both randomization and nondeterminism, but in this order only, first flip as many random bits as necessary and then use nondeterminism. The machine is said to accept a language L if for every string in L the probability that there exists an accepting computation is at least 2/3, and for every string not in L the probability is at most 1/3 (the probability is over all random coin flips, and the existence is over all nondeterministic choices). The class AM is the set of languages accepted by some AM machine that runs in polynomial time. The randomization stage of the computation is called the "Arthur" stage, and the second stage, the nondeterministic one, is called the "Merlin" stage. For exact definitions as well as motivation see [Bab85, BM88, GS86].

In [BM88, GS86] the question of whether $AM = almost - NP$ was raised. This would strengthen the feeling that AM is the probabilistic analogue of NP. Our results imply that this is indeed the case.

THEOREM 11 $AM = almost \quad NP$.

Proof: We first show that $AM \subseteq almost - NP$. Given an AM machine we can first reduce the probability of error such that for a given $\epsilon > 0$, on any input of length n, the machine errs with probability bounded by $\epsilon 4^{-n}$. An NP machine equipped with a random oracle can use the oracle to simulate the *Arthur* phase of the AM machine. For any given input, this machine will accept with the same probability as the AM machine. By summing the probabilities of error over all possible inputs we get that the probability that this machine errs on *any* input is at most ϵ. Since ϵ is arbitrary we get that $AM \subseteq almost - NP$.

We will now prove $almost - NP \subseteq AM$. We first prove the following fact:

Fact: If L $\in$ almost-NP then there exists a specific nondeterministic oracle Turing machine M that runs in polynomial time such that for an oracle A chosen at random:

$$Pr[M^A \text{ accepts } L] \geq 2/3 \tag{2.2.11}$$

Proof of Fact: Since there are only countably many Turing machines, some fixed Turing machine accepts the language L on non-zero measure of oracles. By using the Lebesgue density theorem, we see that it is possible to fix some finite prefix of the oracle such that for oracles with this prefix the Turing machine accepts L with probability at least $2/3$. Finally, this prefix can be hard-wired into the Turing machine.

Up to this point we have only used the standard tools. The difficulty comes when we try to simulate M (with a random oracle) by an AM machine. The difficulty lies in the fact that the machine may access (nondeterministically) an exponential number of locations of the oracle, but AM computations can only supply a polynomial number of random bits. We will use our generator to convert a polynomial number of random bits to an exponential number of bits that "look" random to the machine M.

Let the running time of M be n^k. We can view the computation of M as a large OR of size 2^{n^k} of all the deterministic polynomial time computations occurring for the different nondeterministic choices. Each of these computations can be converted to a DNF formula of size 2^{n^k} over the oracle entries. All together the computation of M can be written as a depth 2 circuit of size at most 2^{2n^k} over the oracle queries.

Our generator can produce from $2n^{10k}$ random bits 2^{2n^k} bits that look random to any depth 2 circuit of this size. So the simulation of M on a random oracle proceeds as follows: Arthur will flip $2n^{10k}$ random bits, and then M will be simulated by Merlin; whenever M makes an oracle query, the answer will be generated from the random bits according to the generator. Note that this is just a parity function of some subset of the bits, which can clearly be computed in polynomial time. Since the generator "fools" this circuit, the simulation will accept with approximately the same probability that M accepts on a random oracle. $\square$

Exactly the same technique suffices to show that for any computation in PH, the polynomial time hierarchy [Sto76, CKS81], a random oracle can be substituted by an "Arthur" phase. Applying to this Sipser's [Sip83] result that $BPP \subseteq \Sigma_2 \cap \Pi_2$ allows simulation of the "Arthur" phase by one more alternation and thus we get:

THEOREM 12 $almost - PH = PH$

2.2.5   BPP and the Polynomial Time Hierarchy

In [Sip83] Sipser showed that BPP could be simulated in the polynomial time hierarchy. Gacs improved this result and showed simulation is possible in $\Sigma_2 \cap \Pi_2$. In this section we give a new simple proof of this fact.

THEOREM 13 (Sipser, Gacs) $BPP \subseteq \Sigma_2 \cap \Pi_2$.

Proof: Since BPP is closed under complement, it suffices to show that $BPP \subseteq \Sigma_2$. The main idea is that a pseudorandom generator that stretches $O(\log n)$ random bits to n pseudorandom bits can be constructed in Σ_2. To simulate BPP then, a Σ_2 machine will then run over all of the polynomially many possibilities of the random seed.

To get such a pseudorandom generator, using our construction, we only need a function with exponential hardness (specifically we want a function on $O(\log n)$ bits with hardness which is $\Omega(n^2)$). Such a function can be found in Σ_2: A simple counting argument shows that such a function exists (although non uniformly), and verifying that a function on $O(\log n)$ bits has indeed a high hardness can easily be seen to be in Co-NP. (The function can be described by a polynomial size table, and the verification can be done by nondeterministically trying all circuits of size n^2).

Thus the simulation will proceed as follows: (1) Nondeterministically guess a function on $O(\log n)$ bits with high hardness (first alternation). (2) Verify it is indeed hard (second alternation). (3) Use it as a basis for the pseudorandom generator, using our construction. (4) Try all possible seeds. $\square$

Actually, this proves a slightly stronger statement, namely that $BPP \subseteq ZPP^{NP}$. (ZPP^{NP} is the class of languages that have polynomial time, randomized, *zero error* algorithms, using an NP-complete oracle).

2.2.6 Randomness and Time vs. Space

Our generator is based on the assumption that there exists a function in, say, $DTIME(2^n)$ that cannot be approximated by small circuits. In this section we show that if this assumption does not hold then some nontrivial simulation of time by space is possible.

This result shows that either randomized algorithms can be simulated deterministically with subexponential penalty, or that, in some sense, an algorithm that runs in time T can be simulated in space $T^{1-\epsilon}$, for some $\epsilon > 0$. This simulation is significantly better than the best known simulation of time T in space $T/\log T$ due to Hopcroft, Paul and Valiant [HPV75]. A result of a similar flavor, giving a tradeoff between simulation of of randomness by determinism and of time by space, was proved using different methods by Sipser [Sip] under an *unproven assumption* regarding certain strong expanders.

Our results are obtained by considering the circuit complexity of the following function: On input $< M, x, t >$ the output is a representation of what Turing Machine M does on input x at time t, where the representation includes the state the machine is in, the location of the heads, and the contents of the cells in which the heads are. Let L be any language that encodes this function, and let L_n be the restriction of L to strings of length n.

Hypothesis $H(\epsilon, n)$: There is a circuit of size $2^{(1-\epsilon)n}$ that computes L_n.

We will show that if hypothesis H is true then some nontrivial simulation of time by space is possible, and if it is false then we can use our construction to get a pseudorandom bit generator.

LEMMA 6 If hypothesis $H(\epsilon, n)$ is true for some $\epsilon > 0$ and all sufficiently large n then for some constants $C > 1$ and $\epsilon > 0$, and for every function $T(n) = \Omega(C^n)$, $DTIME(T(n)) \subseteq DSPACE(T^{1-\epsilon}(n))$.

LEMMA 7 If for every $\epsilon > 0$, hypothesis $H(\epsilon, n)$ is false for all sufficiently large n, then for every $\epsilon > 0$, there exists a pseudorandom generator $G : n^\epsilon \to n$ that is computable in polynomial time.

Proof: (of lemma 6) We will show that (1) if for some $\epsilon > 0$ Hypothesis $H(\epsilon, n)$ is true for all n then $L \in DSPACE(2^{(1-\epsilon)n})$ and that (2) this implies the lemma.

To prove (1) we present a space-efficient algorithm for L: The machine tries all circuits of size $2^{(1-\epsilon)n}$; for each one it checks whether this is indeed the circuit for L. Once it finds the correct circuit, it uses it to look up the answer. Note that checking whether the circuit is the correct one is easy, since it only needs to be consistent between consecutive accesses to the same cell.

To prove (2) consider any Turing machine M running in $DTIME(T(n))$ where $T(n) = 2^{t(n)}$. The result of the Turing machine can be derived by queries to L that encode the value of the function on $< M, x, T(n) >$. This can be done is $DSPACE(2^{(1-\epsilon)m})$ where m is the size of the input which in this case is $n + t(n) + K$, where K is the length of the description of M. It can be easily checked that the statement of the lemma follows. $\square$

Note: Actually a stronger statement can be made, as under the assumption H the simulation mentioned can even be performed in $\Sigma_2 - TIME(T^{(1-\epsilon)}(n))$.

Proof: (of lemma 7) First note that if $H(\epsilon, n)$ is false then every circuit of size $2^{n/2}$ errs on at least $2^{-\epsilon n}$ fraction of the inputs, since otherwise there would be at most $2^{(1-\epsilon)n}$ errors which could be corrected by a table. Next, Yao's Xor lemma (lemma 2) allows amplification of the unpredictability by Xoring disjoint copies of L. By taking $2^{k\epsilon n}$ disjoint copies (for an appropriately chosen constant k), a function with arbitrary large polynomial hardness can be reached, which can be used as the basis for our generator. $\square$

The exact statement of the theorem we obtain is thus:

THEOREM 14 One of the following two possibilities holds:

1. $BPP \subseteq \bigcap_{\epsilon>0} DTIME(2^{n^{\epsilon}})$.

2. There exist $\epsilon > 0$ and $C > 1$ such that for any function $T(n) = \Omega(C^n)$, every language in $DTIME(T(n))$ has an algorithm for it that for infinitely many n, runs in SPACE (actually even $\Sigma_2 - TIME$) $T^{(1-\epsilon)}(n)$ on all inputs of length n.

Proof: If for every $\epsilon > 0$ hypothesis $H(\epsilon, n)$ holds for only finitely many n then lemma 7 assures the existence of pseudorandom generators $G : n^{\epsilon} \rightarrow n$, and by lemma 1 possibility (1) is true. Otherwise the algorithm in the proof of lemma 6 will work for some $\epsilon > 0$ and infinitely many n which implies possibility (2). $\square$

3 Multiparty Protocols and Pseudorandom Generators for Logspace

This chapter describes how to construct pseudorandom generators for Logspace. The main combinatorial tool we use is a certain multiparty communication game.

Section 1 describes the multiparty communication game and proceeds to give lower bounds on the communication complexity of a certain function. We conclude this section by mentioning some results related to our bounds.

In section 2 we define and discuss pseudorandom generators for Logspace. We then show how to use the lower bounds given in section 1 in order to construct them. We finally give some applications of our pseudorandom generators.

3.1 Multiparty Communication Complexity

3.1.1 Definitions

Chandra, Furst and Lipton ([CFL83]) introduced the following multiparty communication game: Let $f(x_1 \ldots x_k)$ be a Boolean function that accepts k arguments each n bits long. There are k parties, each having unlimited computational power, who wish to collaboratively evaluate f. The i'th party knows all the input arguments *except* x_i. They share a blackboard, viewed by all parties, where they can exchange messages. The objective is to minimize the number of bits written on the board.

The game proceeds in rounds. In each round some party writes one bit on the board. The last bit written on the board is considered the outcome of the game and should be the value of $f(x_1 \ldots x_k)$. The protocol specifies which party does the writing and what is written in each round. It must specify the following information for each possible sequence of bits that is written on the board so far:

- Whether the game is over, and in case it is not over, which party writes the next bit: this should be completely determined by the information written on the board so far.
- What that party writes: this should be a function of the information written on the board so far and of the parts of the input that the party knows.

Definition 11 : The *cost* of a protocol is the number of bits written on the board for the worst case input. The *multiparty communication complexity* of f, $C(f)$, is the minimal cost of a protocol that computes f.

We will also be interested in the number of bits needed in order to compute f correctly on even *most* inputs.

Definition 12 : The *bias* a protocol P achieves on f, $B(P, f)$, is defined to be:

$$B(P, f) = |Pr[P(x) = f(x)] - Pr[P(x) \neq f(x)]| \tag{3.1.1}$$

Where $x = (x_1 \ldots x_k)$ is chosen uniformly over all k-tuples of n-bit strings.

Definition 13 : The *ϵ-distributional communication complexity* of f, $C_\epsilon(f)$, is the minimal cost of a protocol which achieves a bias of at least ϵ on f.

Let us just mention that it is possible to define natural probabilistic or nondeterministic analogues of the multiparty communication complexity. Although we will not be concerned with them, the interested reader may notice that our techniques are strong enough to give lower bounds for all these complexity measures. Also, it is possible to consider the *average* complexities, and our techniques suffice to bound the average complexities as well.

3.1.2   Previous Work

For the special case $k = 2$, this multiparty game is exactly the game standard in communication complexity theory, where one party knows x_1, and the other x_2. This case has been extensively investigated in many different contexts and many different lower bounds appear in the literature ([AUY83, Yao79, BFS86] and many more). The distributional communication complexity has been studied as well. Yao [Yao83] first considered the distributional communication complexity and proved a lower bound of $\Omega((\log n)^2)$ for the "inner product mod 2" function. Vazirani [Vaz85] improved this bound to $\Omega(n/\log n)$, and Chor and Goldreich [CG85] improved it to $\Omega(n)$.

For other values of k less is known. Chandra, Furst and Lipton [CFL83] Considered the complexity of the function E_N defined by $E_N(x_1 \ldots x_k) = 1$ iff $x_1 + x_2 + \ldots + x_k = N$. They showed that for $k = 3$, E_N has a communication complexity of $\Omega(\sqrt{\log N})$ and for general k they only succeeded in showing that the complexity is $\omega(1)$. For the distributional communication complexity, no previous lower bounds were known.

3.1.3   Cylinder Intersections

In this subsection we study the basic structure that a multiparty protocol induces on the set of possible inputs, the set of k-tuples.

Consider a multiparty protocol for evaluating a function. For every possible k-tuple $x = (x_1 \ldots x_k)$ a certain communication takes place, and some string is written on the board. The k-tuples may be partitioned according to the string that gets written on the board.

Definition 14 Let s be a string and P a multiparty protocol. The s-component, $X_{P,s}$, is defined to be the set of k-tuples $x \in (\{0,1\}^n)^k$ such that on input x the protocol P results in exactly s being written on the board.

The s-components have a very special structure, which we will now specify.

Definition 15 A subset S of k-tuples is called a *cylinder in the i'th dimension*, if membership in S does not depend on the i'th coordinate. A subset of k-tuples is called a *cylinder intersection* if it can be represented as an intersection of cylinders.

LEMMA 8 For any protocol P and string s, $X_{P,s}$ is a cylinder intersection.

Proof: Define S_i to be the set of k-tuples that is consistent with the communication pattern from the i'th party point of view. I.e.

$$S_i = \{(x_1 \ldots x_k) : \text{ for some } x_i', (x_1, ..., x_i', ..., x_k) \in X_{P,s}\} \tag{3.1.2}$$

It is clear that for each i, S_i is a cylinder in the i'th coordinate. We will show that $X_{P,s} = \bigcap_i S_i$.

It is clear that $X_{P,s} \subseteq \bigcap_i S_i$, it remains to show that $\bigcap_i S_i \subseteq X_{P,s}$. Let $(x_1 \ldots x_k) \in \bigcap_i S_i$, then for every i there exists x_i' such that $(x_1, ..., x_i', ..., x_k) \in X_{P,s}$. We claim that on input $(x_1 \ldots x_k)$ the protocol will still write s on the board. The reason is that all through the communication process the i'th party cannot distinguish between the input of $(x_1 \ldots x_k)$ and the input of $(x_1, ..., x_i', ..., x_k)$. Thus, the bits written on the board will never deviate from s. $\square$

Given a protocol which computes a function f, the value of f must be constant over any single s-component. Our lower bounds will be based upon the fact that for our particular functions f, any cylinder intersection *must* contain approximately the same number of 1's and 0's of the function.

Definition 16 Let $f : (\{0,1\}^n)^k \to \{0,1\}$ be a boolean function. The *discrepancy* of f is

$$\Gamma(f) = \max_S |Pr[f(x) = 1 \text{ and } x \in S] - Pr[f(x) = 0 \text{ and } x \in S]| \tag{3.1.3}$$

where S ranges over all cylinder intersections and x is chosen uniformly over all k tuples.

LEMMA 9 For any function f:

$$C(f) \geq \log_2(1/\Gamma(f)) \tag{3.1.4}$$

and

$$C_\epsilon(f) \geq log_2(\epsilon/\Gamma(f)) \tag{3.1.5}$$

Proof: Consider a protocol P achieving a bias of ϵ on f. We can compute the bias of P on f as the sum of the biases achieved on the different s-components.

$$Bias(P, f) = |Pr[P(x) = f(x)] - Pr[P(x) \neq f(x)]| \leq$$
$$\sum_s |Pr[P(x) = f(x) \text{ and } x \in X_{P,s}] - Pr[P(x) \neq f(x) \text{ and } x \in X_{P,s}]| \qquad (3.1.6)$$

where s ranges over all the possible strings that may be written on the board by the protocol.

For any $x \in X_{P,s}$, $P(x)$ was defined to be the last bit of s, thus

$$|Pr[P(x) = f(x) \text{ and } x \in X_{P,s}] - Pr[P(x) \neq f(x) \text{ and } x \in X_{P,s}]| =$$
$$|Pr[f(x) = 1 \text{ and } x \in X_{P,s}] - Pr[f(x) = 0 \text{ and } x \in X_{P,s}]| \qquad (3.1.7)$$

Since $X_{P,s}$ is a cylinder intersection, we get that the last quantity is bounded from above by $\Gamma(f)$. Thus, if M is the number of different possible strings that may be written on the board by the protocol P, we get that

$$Bias(P, f) \leq M \cdot \Gamma(f) \qquad (3.1.8)$$

The statement of the lemma follows since to produce M different strings requires at least $\log_2 M$ bits. $\square$

3.1.4 A Lower Bound for Generalized Inner Product

In this subsection we prove a lower bound on the multiparty communication complexity of the generalized inner product.

Definition 17 The $k - wise\ generalized\ inner\ product$ function on k n-bit strings is defined by $GIP_{n,k}(x_1 \ldots x_k) = 1$ if the number of locations in which all of the x_i's have 1 is odd, and 0 otherwise.

We will prove a lower bound on the communication complexity of GIP by giving an upper bound to the discrepancy. We first introduce a slightly modified notation to facilitate easier algebraic handling.

Definition 18 $f(x_1 \ldots x_k)$ is 1 if $GIP(x_1 \ldots x_k) = 0$ and -1 if $GIP(x_1 \ldots x_k) = 1$.

Definition 19 :

$$\Delta^k(n) = \max_{\phi_1, \ldots, \phi_k} |E_{x_1 \ldots x_k} f(x_1 \ldots x_k) \phi_1(x_1 \ldots x_k) \ldots \phi_k(x_1 \ldots x_k)| \qquad (3.1.9)$$

Where the maximum is taken over all functions $\phi_i : (\{0,1\}^n)^k \to \{0,1\}$ s.t. ϕ_i does not depend on x_i. The E stands for expected value over all the possible 2^{nk} choices of $x_1 \ldots x_k$.

Note that $\Delta^k(n)$ is exactly $\Gamma(GIP_{k,n})$, the discrepancy of the k-wise inner product function on n-bits.

LEMMA 10 Defines constants μ_k by the recursion: $\mu_1 = 0$, and $\mu_k = \sqrt{\frac{1+\mu_{k-1}}{2}}$. Then

$$\Delta^k(n) \le (\mu_k)^n \qquad (3.1.10)$$

Note: It can be shown by induction that $\mu_k \le 1 - 4^{1-k}$ which is approximately $e^{-4^{1-k}}$
Proof: We proceed by induction on k. It is clear that $\Delta^1(n) = 0$, except for the case $n = 0$, where we get 1 (let us define $0^0 = 1$ for this paper).

Let $k \ge 2$, and fix $\phi_1, ..., \phi_k$ that achieve $\Delta^k(n)$. Since ϕ_k does not depend on x_k, and is bounded in absolute value by 1,

$$\Delta^k(n) \le E_{x_1 \ldots x_{k-1}} |E_{x_k} f(x_1 \ldots x_k)\phi_1(x_1 \ldots x_k) \ldots \phi_{k-1}(x_1 \ldots x_k)| \qquad (3.1.11)$$

In order to estimate the right-hand side, we will use the special case of the Cauchy-Schwartz inequality stating that for any random variable z: $(E[z])^2 \le E[z^2]$.

Thus our estimate is:

$$\Delta^k(n) \le [E_{x_1 \ldots x_{k-1}}[E_{x_k} f(x_1 \ldots x_k)\phi_1(x_1 \ldots x_k) \ldots \phi_{k-1}(x_1 \ldots x_k)]^2]^{1/2} =$$
$$[E_{u,v,x_1 \ldots x_{k-1}} f(x_1 \ldots x_{k-1}, u) f(x_1 \ldots x_{k-1}, v)\phi_1^u \phi_1^v \ldots \phi_{k-1}^u \phi_{k-1}^v]^{1/2} \qquad (3.1.12)$$

where ϕ_i^u stands for $\phi_i(x_1 \ldots x_{k-1}, u)$, and ϕ_i^v for $\phi_i(x_1 \ldots x_{k-1}, v)$.

To estimate this we will need the following observation: For every particular choice of u and v, we can express $f(x_1 \ldots x_{k-1}, u) f(x_1 \ldots x_{k-1}, v)$ in terms of the function f on $k-1$ strings of a possibly shorter length. Inspection reveals that the value of $f(x_1 \ldots x_{k-1}, u) f(x_1 \ldots x_{k-1}, v)$ is simply $f(z_1 \ldots z_{k-1})$ where z_i is the restriction of x_i to the coordinates j such that $u_j \ne v_j$. We will now view each x_i as composed of two parts: z_i and y_i, where z_i is the part of the x where $u_j \ne v_j$, and y_i the rest (this is done separately for every u, v).

For every particular choice of u, v and consequently $y_1 \ldots y_{k-1}$, we define functions of the "z-parts":

$$\xi_i^{u,v,y_1 \ldots y_{k-1}}(z_1 \ldots z_{k-1}) = \phi_i(x_1 \ldots x_{k-1}, u)\phi_i(x_1 \ldots x_{k-1}, v) \qquad (3.1.13)$$

where the x_i's are obtained by the concatenation of the corresponding y_i and z_i. We can now rewrite the previous estimate as

$$\Delta^k(n) \leq [E_{u,v} E_{y_1 \ldots y_{k-1}} S^{u,v,y_1 \cdots y_{k-1}}]^{1/2} \qquad (3.1.14)$$

where $S^{u,v,y_1 \cdots y_{k-1}}$ is defined by

$$S^{u,v,y_1 \cdots y_{k-1}} =$$
$$E_{z_1 \ldots z_{k-1}} f(z_1 \ldots z_{k-1}) \xi_1^{u,v,y_1 \cdots y_{k-1}}(z_1 \ldots z_{k-1}) \ldots \xi_{k-1}^{u,v,y_1 \cdots y_{k-1}}(z_1 \ldots z_{k-1}) \qquad (3.1.15)$$

Now, $S^{u,v,y_1 \cdots y_{k-1}}$ can be estimated via the induction hypothesis. Indeed note that $\xi_i^{u,v,y_1 \cdots y_{k-1}}$ does not depend on z_i. Thus the previous estimate of $\Delta^k(n)$ is bounded by

$$\Delta^k(n) \leq [E_{u,v,y_1 \cdots y_{k-1}} \Delta^{k-1}(m_{u,v})]^{1/2} \leq [E_{u,v,y_1 \cdots y_{k-1}} \mu_{k-1}^{m_{u,v}}]^{1/2} \qquad (3.1.16)$$

Where $m_{u,v}$ is the length of the strings z_i, which is equal to the number of locations j such that $u_j \neq v_j$.

Since u and v are distributed uniformly in $\{0,1\}^n$, $m_{u,v}$ is distributed according to the binomial distribution. For any constant m, the probability that $m_{u,v} = m$ is exactly $\binom{n}{m} 2^{-n}$. Thus the previous estimate is given by:

$$\Delta^k(n) \leq [\sum_{m=0}^{n} \binom{n}{m} 2^{-n} \mu_{k-1}^m]^{1/2} = [2^{-n}(1 + \mu_{k-1})^n]^{1/2} = \mu_k^n \qquad (3.1.17)$$

Which completes the proof of the lemma. $\square$

Combining this bound with lemma 2.4 we get:

THEOREM 15 $C_\epsilon(GIP_{k,n}) = \Omega(n/4^k + \log_2 \epsilon)$

3.1.5 Further Results

Our techniques and bounds have several other applications and implications. Since these results are not directly related to pseudorandom generation, we will only describe them briefly here. Complete definitions, discussion and proofs appear in [BNS89].

Other lower bounds Lower bounds on the multiparty communication complexity can also be proved for other functions. In particular, bounds which are slightly stronger than in Theorem 15 are proven for the following function:

Definition 20 The *quadratic character of the sum mod p* function is: $QC_{p,k}(x_1 \ldots x_k) = 1$ iff $x_1 + \ldots + x_k$ is a quadratic residue mod p.

THEOREM 16 For any n-bit long prime number p: $C_\epsilon(QC_{p,k}) = \Omega(n/2^k + \log_2 \epsilon)$

Time-Space tradeoffs for Turing machines Multiparty protocols are a general model of computation, and lower bounds on their complexity may be used to obtain lower bounds in other models. Our lower bounds for the generalized inner product function imply the following time-space tradeoff for general Turing machines:

THEOREM 17 Any k-head Turing machine computing the $(k+1)$-wise generalized inner product function on n-bit strings requires a time-space tradeoff of $TS = \Omega(n^2)$.

This bound is tight, and significantly better than the previously known tradeoffs for k-head Turing machines [Kar86, GS88, DG84].

Branching programs Chandra, Furst and Lipton [CFL83] observe how lower bounds for Multiparty protocols can be used to prove length-width tradeoffs for branching programs. Using different techniques, related to those used in [AM86], and relying on our lower bounds for multiparty communication complexity, it is possible to obtain length-width tradeoffs for branching programs. These bounds also imply new lower bounds on the size of branching programs and on the size of boolean formula computing certain functions.

3.2 Pseudorandom Generators for Logspace

In this section we show how to use our lower bounds for multiparty protocols in order to construct pseudorandom generators for Logspace (or generally any small-space machines).

3.2.1 On Randomized Space

There are quite a few subtle points to consider when defining randomized space bounded complexity classes. We will present here the "correct" definition. For an overview of the subtleties involved refer, e.g., to [BCD$^+$89].

Definition 21 A *randomized, space $s(n)$* Turing machine is a coin flipping Turing machine such that:

1. The Turing machine runs in space $s(n)$ on any input of size n.
2. The Turing machine may flip a fair unbiased coin at any stage.
3. The Turing machine may never get into an infinite loop, for any sequence of coin flips. In particular, with probability 1 it terminates in time $exp(s(n))$ on any input of length n.

A Turing machine accepts a language L with one-sided error if for every $x \in L$ the machine accepts with probability of at least $1/2$ and for any x not in L it rejects with probability 1. A Turing machine accepts a language L with two-sided error if for any $x \in L$, the Turing machine accepts with probability of at least $2/3$, and for any $x \notin L$ it rejects with probability of at least $2/3$.

$RSPACE(s(n))$ is the class of languages accepted with one-sided error by a space $s(n)$ randomized Turing machine. $BPSPACE(s(n))$ is the class of languages accepted with two-sided error by a randomized space $s(n)$ Turing machine. RL is $RSPACE(O(\log n))$, and BPL is $BPSPACE(O(\log n))$.

We want to focus attention on condition (2), the kind of access the machine has to the random bits. As defined, the randomized Turing machine has access to the random bits one by one. When it wants a random bit it can flip a coin, but in no case can it go "back" and review the result of a previous coin flip. Any bit that it wishes to "remember" must be kept in the limited storage. This restriction that the machine does not have multiple, 2-way, access to the random bits is essential, as, for example, random-Logspace machines with 2-way access to the random bits are not even known to be in deterministic polynomial time (in contrast to RL and BPL being in P).

It is interesting to note that the same apparent difference in power between one-way and two-way access holds also for nondeterministic computation. In the "correct" definition of NL, the Turing machine has one-way access to the nondeterministic bits. If the definition is changed to allow two-way access to nondeterministic bits, then the machines turn out to have the full power of NP.

3.2.2 Space Bounded Statistical Tests

We are interested in producing pseudorandom sequences that might be used instead of truly random sequences in space bounded computations. Thus the statistical tests that must be passed by the generator are the ones that can be performed by a space bounded Turing machine on its *random source*. Note that this class of tests may be a proper subset of the tests that may be performed by space bounded machines on their *input tape*.

We will allow non-uniform statistical tests as well.

Definition 22 A *space-$s(n)$ statistical test* is a deterministic space $s(n)$ Turing machine M, and an infinite sequence of binary strings $a = (a_1, ..., a_n, ...)$ called the advice strings. We require that the length of a_n is at most $exp(s(n))$.

The result of the test on input x, $M^a(x)$, is determined as follows: The string a_n, where n is the length of x, is put on a special read-only tape of M called the advice tape. The machine M is run on the advice tape, treating it as a normal input tape. The machine has the following one-way mechanism to access x: at any point it may request the next bit of x (in the same fashion that a randomized Turing machine may request the next random bit).

Notice that these tests have considerable power. We next give some examples of operations that may be performed and combined by Logspace tests:

Consider the input as partitioned into consecutive words, each of some fixed small length ($O(\log n)$). The following questions may all be answered by a Logspace test.

- Count the number of times a certain value appears.
- Compute the average value of a word.
- Compute the standard deviation, or higher moments.
- Compute any of the previous measures for an arbitrary subset of the words, or of the bits.

In fact the vast majority of the statistical tests described by Knuth [Knu] lie in this class.

3.2.3   Definition of Pseudorandom Generators for Space Bounded Computation

A Pseudorandom generator for space $s(n)$ must produce strings that look random to any space $s(n)$ statistical test.

Definition 23 A function $G = \{G_n : \{0,1\}^{l(n)} \rightarrow \{0,1\}^n\}$ is called a *pseudorandom generator for space* $s(n)$, if for every polynomial $p(n)$, all large enough n, and every space $s(n)$ statistical test M^a,

$$|Pr[M^a(y)\ accepts] - Pr[M^a(G(x))\ accepts]| \leq 1/p(n) \tag{3.2.18}$$

where y is chosen uniformly in $\{0,1\}^n$, and x uniformly in $\{0,1\}^{l(n)}$.

The first observation we should make when trying to construct a pseudorandom generator for space bounded Turing machines is that without loss of generality we can restrict the class of statistical tests that have to be passed. In a similar fashion to Yao's [Yao82] results for general (polytime-hard) pseudorandom generators, we show that any generator that passes all space $s(n)$ *prediction* tests will be a pseudorandom generator for space $s(n)$.

Definition 24 A function $G = \{G_n : \{0,1\}^{l(n)} \rightarrow \{0,1\}^n\}$ *passes all space* $s(n)$ *prediction tests to within* $\epsilon(n)$ if for every polynomial $p(n)$, every space $s(n)$ statistical test M^a, and every $1 \leq i \leq n$,

$$|Pr[M^a(first\ i-1\ bits\ of\ G(x)) = i'th\ bit\ of\ G(x)] - \frac{1}{2}| \le \epsilon(n) \qquad (3.2.19)$$

where x is chosen uniformly in $\{0,1\}^{l(n)}$. G is said to pass all *space-$s(n)$* prediction tests if for every polynomial $p(n)$, G passes all *space-$s(n)$* prediction tests to within $1/p(n)$ for all sufficiently large n.

LEMMA 11 G is a pseudorandom generator for space $s(n)$ iff it passes all space $s(n)$ prediction tests.

Proof: Similar to Yao's proof. $\square$

3.2.4 Description of the Generator

Our generator is based on a function f that takes k arguments each r bits long, and has high multiparty communication complexity.

Definition 25 The *input* to the generator will consist of t boolean strings, each r bits long (all together rt random bits). It will produce $\binom{t}{k}$ *output* bits. Each output bit will be of the form $f(S)$ where S is some cardinality-k subset of the input strings. The order is extremely important, and we will take *all* the k-subsets in the *antilexicographic* order. (I.e. $f(S_1)$ appears before $f(S_2)$ if the latest string in the symmetric difference of S_1 and S_2 is in S_2.)

LEMMA 12 For any $\epsilon > 0$ and $s < C_\epsilon(f)/k$, the above generator passes all space s prediction tests to within ϵ.

Proof: Assume not; we shall give a multiparty protocol that predicts f with bias ϵ using less than $C_\epsilon(f)$ bits of communication. Suppose the l'th bit, $f(S)$, can be predicted by the Turing machine, where $S = \{x_{i_1}, x_{i_2}, ..., x_{i_k}\}$, and $i_1 > i_2 > ... > i_k$. We shall now show how k parties, the j'th knowing the values of each x_i in S *except* x_{i_j}, can predict $f(x_{i_1}, x_{i_2}, ..., x_{i_k})$, with low communication.

By an averaging argument, it is possible to fix the values of all x_i's not in S to constants in some way, while preserving the prediction bias of the Turing machine. Thus we can assume w.l.o.g that all x_i's not in S are fixed and known to all the parties beforehand.

The parties will simulate the Turing machine running on the first $l-1$ bits of the output of the generator. Since the parties have unlimited power, they have no problem simulating the Turing machine as long as they have access to the input bits which the Turing machine reads (or can compute them from the information they have). In our case, however, no single party can do so since each party is missing the value of x_{i_j} for some j, which is needed to compute some of the input bits (i.e. bits of the form $f(T)$ where T contains x_{i_j}).

In our simulation each party simulates the Turing machine for as long as it can and then sends the current state of the machine to the next party to continue the simulation. The simulation starts with the first party (i.e. the party missing only x_{i_1}). Note that this player can simulate the Turing machine until the first time that a bit involving x_{i_1} is read by the Turing machine. At this point the first player sends the total state of the Turing machine to the second player who continues with the simulation. This player can now continue the simulation until the first point where a bit involving x_{i_2} is read, at which point he sends the state of the machine to the third player, etc. This continues all way to the k'th player.

The important thing to notice is that once the last player starts with the simulation, he can continue it until the prediction of $f(S)$. This is so because of the ordering of the bits of the generator that we chose. Consider the point where the first party cannot continue the simulation since a bit involving x_{i_1} first appears. Because the sets used to compute the bits of the generator were ordered in antilexicographic order, we observe that all sets from this point on, until S contain x_{i_1}. Similarly once x_{i_2} appears it remains in every set until S, etc. So, when the k'th party cannot continue the simulation, the next set must contain all of $x_{i_1}, ..., x_{i_k}$, i.e. it must be S, that is the point where the Turing machine makes the prediction.

Sending the total state of the Turing machine requires s bits, and since this is done in our simulation only $k-1$ times, we obtain a protocol that requires only $(k-1)s$ bits and predicts $f(S)$ with a bias of ϵ. This contradicts the choice of $s < C_\epsilon(f)/k$. $\Box$

We can now state the main result of this section:

THEOREM 18 For some constants $c_1, c_2 > 0$, there exists (an explicit) pseudorandom generator, $G = \{G_n : \{0,1\}^{l(n)} \to \{0,1\}^n\}$, for space $2^{c_1\sqrt{\log n}}$, where $l(n) = 2^{c_2\sqrt{\log n}}$. Moreover, G can be computed by a Logspace Turing machine (having multiple access to its input bits).

Proof: We use the construction of the generator described in this section, with f being the "generalized inner product" function, $k = 2\sqrt{\log n}$, $t = 2^k$, and $r = 8^k$. Theorem 15 guarantees the high communication complexity of this function, and thus lemma 12 shows that the generator passes all prediction tests, and using lemma 11 we conclude that it is a pseudorandom generator. $\Box$

3.2.5   One-way vs. Two-way Access to Randomness

Our generator sheds some light on the difference between one-way and two-way access to the random bits given to Logspace machines. We show that 2-way access is better, at least in the sense that fewer random bits are necessary.

COROLLARY 3 A randomized Logspace Turing machine with one-way access to the random bits that uses a polynomial number of random bits may be simulated by a randomized Logspace machine with 2-way access to the random bits that uses only $2^{O(\sqrt{\log n})}$ random bits.

Proof: Our generator can be implemented by a Logspace machine having 2-way access to the random bits. The generator can be easily run "on the fly", and generate one bit at a time to supply to the original, simulated machine, whenever it flips a coin. $\Box$

3.2.6 Universal Sequences

One of the most interesting subclasses of Logspace statistical tests are those related to walks on graphs. Given a graph H in the advice tape, a Logspace machine can treat the input to the test as directions for a walk on the graph and perform the walk. Thus the output of a pseudorandom generator for Logspace will behave like a random walk on any graph. We use this fact in order to construct universal traversal sequences first defined in [AKL+79].

Definition 26 A graph is called (d, n)-*labeled* if it is a d-regular graph on n vertices and the edges adjacent to each vertex are labeled by a permutation of $\{1, ..., d\}$ (an edge may be labeled differently at each of its 2 vertices). A string $w \in \{1, ..., d\}^*$ is said to *cover* an (n, d)-labeled graph, if w, when treated as directions to a walk on the graph, visits all vertices of the graph, whatever the starting vertex is.

A string $w \in \{1, ..., d\}^*$ is said to be a (n, d) *universal traversal sequence* if it covers every (d, n)-labeled graph.

In [AKL+79] it is shown that a random string of length $O(d^2 n^3 \log n)$ is a (d, n)-universal sequence with high probability. Results shown in [KLNS88] imply that a random string of length $O(dn^3 \log n)$ actually suffices. However, explicit construction of short universal sequences is more difficult. Explicit constructions are known for two special cases: for $d = 2$ an explicit construction is known of polynomial length universal sequences [Ist88]; and for $d = n$ an explicit construction of length $n^{\log n}$ is known [KPS88]. Our pseudorandom sequences allow us to give (d, n)-universal sequences of length $2^{2^{O(\sqrt{\log n})}}$ for *all* values of d.

We will be using the output of the generator as a random walk. A technical issue that should be mentioned is that we need to convert the *binary* string which is the output of the generator to a string in $\{1, ..., d\}$. One way to do this is to take, say, every $5 \log n$ bits consecutive bits modulo d. This way a uniform distribution on binary string will be converted to an almost uniform distribution on walks.

LEMMA 13 Let $G = \{G_n : \{0,1\}^{l(n)} \to \{0,1\}^{n^4}\}$ be a pseudorandom generator for Logspace. Then for every (d, n)-labeled graph H, $G(x)$ (converted as mentioned) will cover H with probability at least $1/2$ (probability taken over a random choice of x).

Proof: The description of H can be put in the oracle, and then a Logspace machine can perform the walk on H given by its input. A truly random string y of length n^4 converted in this manner will result in a nearly uniformly random walk on H, and as such will visit every vertex, starting from every vertex with probability of at least $1 - 1/3n^2$. A Logspace machine can determine whether vertex i is reached in a walk starting from vertex j, and thus for every i, j, the probability that $G(x)$ (converted to a walk) will reach vertex i starting from vertex j, should be at least $1 - 1/2n^2$. Thus the probability that there exist i, j such that the walk from i does not reach j is at most $1/2$. $\square$

LEMMA 14 Let $G = \{G_n : \{0,1\}^{l(n)} \to \{0,1\}^{n^4}\}$ be a pseudorandom generator for Logspace. Then the string achieved by the concatenation of $G(x)$ for all possible $2^{l(n)}$ values of x (and converted as mentioned) is a (d, n) universal traversal sequence.

Proof: For each (n, d)-labeled graph H, half the substrings of the form $G(x)$ will cover H. Thus when one of these substrings is reached, whatever vertex the walk is in, H will be covered by it. $\square$

Applying our generator to this lemma we get:

THEOREM 19 For every d and n, there exists an (explicitly given) (d, n) universal traversal sequence of length $2^{2^{O(\sqrt{\log n})}}$. Moreover the sequence can be constructed by a Turing machine running in space logarithmic in the length of the sequence.

4 Recent Results

This chapter presents a short survey of results related to the subject matter of this thesis that were obtained only after this thesis was originally written.

4.1 Approximating vs. Computing

The kind of assumptions needed for the pseudorandom generators constructed in chapter two are the existence of a function that cannot even be *approximated* by small circuits. Recently there has been much research based upon the fact that certain functions have "random-self-reducibility" properties, and thus if they can even be approximated then they can be computed.

These ideas have been used by Babai, Fortnow, Nisan and Wigderson [BFNW91] to weaken the assumptions required for the generators of chapter two to the existence of functions that cannot be *computed* by small circuits. Many of the corollaries presented in chapter two thus become stronger.

4.2 Pseudorandom Generators for AC^0

Although no new results regarding pseudorandom generators for constant depth circuits were obtained, an interesting conjecture related to this was studied. Linial and Nisan [LN90], and independently Luby and Velickovic [LV91] conjecture that any generator that produces polylog-wise independent random bits is a pseudorandom generator for constant depth circuits. Some progress regarding this question appears in [LN90, LMN89, LV91].

4.3 Approximating $\#DNF$

Luby and Velickovic [LV91] obtained some progress on the problem of approximating the number of solutions of DNF formulas.

4.4 Multiparty Protocols

Goldman and Hastad [HG90] used our lower bounds for multiparty protocols to prove lower bounds for certain types of depth three circuits that use majority gates. Their technique also shows that proving a lower bound for the multiparty communication complexity for a super-polylogarithmic number of players will suffice to give lower bounds for ACC circuits.

Combining these results with Yao's recent simulations of ACC circuits [Yao90] shows that for $k = \omega(\log n)$ the communication complexity of GIP_k is indeed low.

4.5   Pseudorandom Generators for Space-Bounded Computation

In [Nis90] a different construction for pseudorandom generators for Logspace is presented. This construction uses fewer random bits than the one obtained here in chapter three does. The applications regarding universal sequences are improved similarly.

In [Nis91b] these new pseudorandom generators are used in order to give an SC (=polynomial time, polylogarithmic space) simulation of randomized Logspace.

Bibliography

[AH87]   L.M. Adelman and M.A. Huang. Recognizing primes in random polynomial time. In *Proceedings of the 19th Annual ACM Symposium on Theory of Computing, New York City*, 1987.

[AKL$^+$79] R. Aleliunas, R.M. Karp, R.J. Lipton, L. Lovasz, and C. Rackoff. Random walks, universal sequences and the complexity of maze problems. In *20th Annual Symposium on Foundations of Computer Science, San Juan, Puerto Rico*, 1979.

[AM86]   N. Alon and W. Mass. Meanders, Ramsey theory and lower bounds for branching programs. In *27th Annual Symposium on Foundations of Computer Science, Toronto, Ontario, Canada*, 1986.

[AUY83]   A.V. Aho, J.D. Ullman, and M. Yanakakis. On notions of information transfer in VLSI circuits. In *Proceedings of the 15th Annual ACM Symposium on Theory of Computing, Boston, Massachusetts*, 1983.

[AW85]   M. Ajtai and A. Wigderson. Deterministic simulation of probabilistic constant depth circuits. In *26th Annual Symposium on Foundations of Computer Science, Portland, Oregon*, pages 11–19, October 1985.

[Bab85]   L. Babai. Trading group theory for randomness. In *Proceedings of the 17th Annual ACM Symposium on Theory of Computing, Providence, Rhode Island*, pages 421–429, 1985.

[BCD$^+$89] A. Borodin, S.A. Cook, P.W. Dymond, W.L. Ruzzo, and M. Tompa. Two applications of inductive counting for complementation problems. *SIAM J. Comput.*, 18(3):559–578, 1989.

[BFNW91]   L. Babai, L. Fortnow, N. Nisan, and A. Wigderson. BPP has weak subexponential simulations unless EXPTIME has publishable proofs. In *Proceedings of Structure in Complexity Theory*, 1991.

[BFS86]   L. Babai, F. Frankl, and J. Simon. Complexity classes in communication complexity theory. In *27th Annual Symposium on Foundations of Computer Science, Toronto, Ontario, Canada*, 1986.

[BG81]   C.H. Bennett and J. Gill. Relative to a random oracle A, $P^A \neq NP^A \neq Co - NP^A$ with probability 1. *SIAM J. Comput.*, 10:96–113, 1981.

[BM84]   M. Blum and S. Micali. How to generate cryptographically strong sequences of pseudorandom bits. *J. SIAM*, 13(4):850–864, November 1984.

[BM88]   L. Babai and S. Moran. Arthur Merlin games: a randomized proof system, and a hierarchy of complexity classes. *J. Comp. and Syst. Sci.*, 36(2):254–276, 1988.

[BNS89]   L. Babai, N. Nisan, and M. Szegedy. Multiparty protocols and Logspace-hard pseudorandom sequences. In *Proceedings of the 21st Annual ACM Symposium on Theory of Computing, Seattle, Washington*, pages 1–11, 1989.

[CFL83]   A. Chandra, M. Furst, and R.J. Lipton. Multiparty protocols. In *24th Annual Symposium on Foundations of Computer Science, Tucson, Arizona*, 1983.

[CG85]   B. Chor and O. Goldreich. Unbiased bits from weak sources of randomness. In *26th Annual Symposium on Foundations of Computer Science, Portland, Oregon*, 1985.

[CKS81]   A. Chandra, D. Kozen, and L. Stockmeyer. Alternation. *J. ACM*, 28, 1981.

[DG84]   P. Duris and Z. Galil. A time-space tradeoff for language recognition. *Math. Systems Theory*, 17:3–12, 1984.

[GM84]   S. Goldwasser and S. Micali. Probabilistic encryption. *J. Comp. and Syst. Sci.*, 28(2), 1984.

[GMW86]   O. Goldreich, S. Micali, and A. Wigderson. Proofs that yield nothing but their validity and a methodology of cryptographic protocol design. In *27th Annual Symposium on Foundations of Computer Science, Toronto, Ontario, Canada*, 1986.

[GS86]   S. Goldwasser and M. Sipser. Private coins versus public coins in interactive proof systems. In *Proceedings of the 18th Annual ACM Symposium on Theory of Computing, Berkeley, California*, pages 59–68, May 1986.

[GS88]   Y. Gurevich and S. Selah. Nondeterministic linear tasks may require substantially nonlinear deterministic time in the case of sublinear work space. In *Proceedings of the 20th Annual ACM Symposium on Theory of Computing*, 1988.

[Has86]   J. Hastad. *Computational limitations for small depth circuits*. MIT Press, 1986. Ph.D. thesis.

[Has90]   J. Hastad. Pseudorandom generators under uniform assumptions. In *Proceedings of the 22st Annual ACM Symposium on Theory of Computing, Baltimore, Maryland*, pages 395–404, 1990.

[HG90]   J. Hastad and M. Goldman. On the power of small depth threshold circuits. In *focs31*, pages 610–618, 1990.

[HPV75]   J. Hopcroft, W. Paul, and L. Valiant. On time versus space and related problems. In *16th Annual Symposium on Foundations of Computer Science, Berkeley, California*, 1975.

[ILL89]   R. Impagliazzo, L. Levin, and M. Luby. Pseudorandom generation from one-way functions. In *Proceedings of the 21st Annual ACM Symposium on Theory of Computing, Seattle, Washington*, 1989.

[Ist88]   S. Istrail. Polynomial traversing sequences for cycles are constructable. In *Proceedings of the 20th Annual ACM Symposium on Theory of Computing*, 1988.

[Kar86]   M. Karchmer. Two time-space tradeoffs for element distinctness. *Theor. Comp. Sci.*, 47, 1986.

[KL83]   R.M. Karp and M. Luby. Monte-carlo algorithms for enumeration and reliability problems. In *24th Annual Symposium on Foundations of Computer Science, Tucson, Arizona*, pages 56–64, 1983.

[KLNS88]   J.D. Kahn, N. Linial, N. Nisan, and M.E. Saks. On the cover time of random walks in graphs. *Journal of Theoretical Probability*, 2(1), 1988.

[Knu]   D. Knuth. *The Art of Computer Programming, Volume II*.

[KPS88]   H. Karloff, R. Paturi, and J. Simon. Universal sequences of length $n^{O(\log n)}$ for cliques. *Info. Proc. Lett.*, 28:241–243, 1988.

[Kur87]   S.A. Kurtz. A note on randomized polynomial time. *SIAM Journal on Computing*, 16(5), 1987.

[KUW86]   R.M. Karp, E. Upfal, and A. Wigderson. Construction a maximum matching is in random NC. *Combinatorica*, 6(1):35–48, 1986.

[LMN89]   N. Linial, Y. Mansour, and N. Nisan. Constant depth circuits, Fourier transform and learnability. In *30th Annual Symposium on Foundations of Computer Science, Reseach Triangle Park, NC*, pages 574–579, October 1989.

[LN90]   N. Linial and N. Nisan. Approximate inclusion-exclusion. *Combinatorica*, 10(4):349–365, 1990.

[LR86]   M. Luby and C. Rackoff. Pseudorandom permutation generators and cryptographic composition. In *Proceedings of the 18th Annual ACM Symposium on Theory of Computing, Berkeley, California*, 1986.

[LV91]   M. Luby and B. Velickovic. On deterministic approximation of dnf. In *Proceedings of the 23rd Annual ACM Symposium on Theory of Computing, New Orleans, Louisiana*, pages 430–438, 1991.

[MVV87]   K. Mulmuley, U.V. Vazirani, and V.V. Vazirani. Matching is as easy as matrix inversion. In *Proceedings of the 19th Annual ACM Symposium on Theory of Computing, New York City*, 1987.

[Nis90]   N. Nisan. Pseudorandom generators for space-bounded computation. In *Proceedings of the 22st Annual ACM Symposium on Theory of Computing, Baltimore, Maryland*, pages 204–212, May 1990.

[Nis91a]   N. Nisan. Pseudorandom bits for constant depth circuits. *Combinatorica*, 11(1):63–70, 1991.

[Nis91b]   N. Nisan. $RL \subseteq SC$. Manuscript, 1991.

[NW88]   N. Nisan and A. Wigderson. Hardness vs. randomness. In *29th Annual Symposium on Foundations of Computer Science, White Plains, New York*, pages 2–12, October 1988.

[Rab80]   M.O. Rabin. Probabilistic algorithm for testing primality. *J. of Number Theory*, 12:128–138, 1980.

[RT84]   J.H. Reif and J.D. Tygar. Towards a theory of parallel randomized computation. TR-07-84, Aiken computation lab., Harvard university, 1984.

[Sha81]   A. Shamir. On the generation of cryptographically strong pseudorandom sequences. In *8th ICALP, Lecture notes in Comp. Sci. 62, Springer-Verleg*, pages 544–550, 1981.

[Sip]   M. Sipser. Expanders, randomness or time vs. space. In *Structure in Complexity Theory, Lecture notes in Computer Science, No. 223.*

[Sip83]   M. Sipser. A complexity theoretic approach to randomness. In *Proceedings of the 15th Annual ACM Symposium on Theory of Computing, Boston, Massachusetts*, pages 330–335, May 1983.

[SS77]   R. Solovay and V. Strassen. A fast monte-carlo test for primality. *SIAM Journal on Computing*, 6:84–85, 1977.

[Sto76]   L. Stockmeyer. The polynomial time hierarchy. *Theor. Comp. Sci.*, 3(1), 1976.

[Vaz85]   U.V. Vazirani. Towards a strong communication complexity theory or generating quasirandom sequences from two communicating semirandom sources. In *Proceedings of the 17th Annual ACM Symposium on Theory of Computing, Providence, Rhode Island*, 1985.

[Yao79]   A. C. Yao. Some complexity questions related to distributed computing. In *Proceedings of the 11th Annual ACM Symposium on Theory of Computing, Atlanta, Georgia*, 1979.

[Yao82]   A. C. Yao. Theory and applications of trapdoor functions. In *23th Annual Symposium on Foundations of Computer Science*, pages 80–91, October 1982.

[Yao83]   A. C. Yao. Lower bounds by probabilistic arguments. In *Proceedings of the 15th Annual ACM Symposium on Theory of Computing, Boston, Massachusetts*, 1983.

[Yao90]   A. C. Yao. On ACC and threshold circuits. In *31st Annual Symposium on Foundations of Computer Science, St. Louis, Missouri*, pages 619–627, 1990.

The MIT Press, with Peter Denning as general consulting editor, publishes computer science books in the following series:

ACL-MIT Press Series in Natural Language Processing
Aravind K. Joshi, Karen Sparck Jones, and Mark Y. Liberman, editors

ACM Doctoral Dissertation Award and Distinguished Dissertation Series

Artificial Intelligence
Patrick Winston, founding editor
J. Michael Brady, Daniel G. Bobrow, and Randall Davis, editors

Charles Babbage Institute Reprint Series for the History of Computing
Martin Campbell-Kelly, editor

Computer Systems
Herb Schwetman, editor

Explorations with Logo
E. Paul Goldenberg, editor

Foundations of Computing
Michael Garey and Albert Meyer, editors

History of Computing
I. Bernard Cohen and William Aspray, editors

Logic Programming
Ehud Shapiro, editor; Fernando Pereira, Koichi Furukawa, Jean-Louis Lassez, and David H. D. Warren, associate editors

The MIT Press Electrical Engineering and Computer Science Series

Research Monographs in Parallel and Distributed Processing
Christopher Jesshope and David Klappholz, editors

Scientific and Engineering Computation
Janusz Kowalik, editor

Technical Communication and Information Systems
Ed Barrett, editor